Characteristics of Acids

Before people used modern, atomic ways of defining a what an acid is, the Greeks, specifically Arrhenius, used a different method. Arrhenius defined an acid as any material that tastes sour and corrodes metal. Acids will turn litmus paper yellow, gradually turning to red with strength. Acids will corrode metals, and neutralize a base, only if the base is not stronger than the acid. Acids will turn phenolphthalein solution entirely transparent. Acids have a sour taste, with strength, it will eventually burn your mouth. Some acids you could find in your house include vinegar, lemon juice, carbonic acid in carbonated beverages and some tannic acid in tea. When your mouth feels leathery, dry or shriveled up after consuming some stout tea, the tannic acid is literally tanning your mouth, the way a leatherworker makes leather. (effects are only temporary.)

Characteristics of Bases

Arrhenius defined bases with the behavior of bitter and soapy. Bases turn litmus paper blue, and as the base gets stronger the paper will progress to a dark purple.

Bases react to phenolphthalein solution by turning it a translucent pink. Proteins are constructed with amino acids, when a base is introduced to the protein, the amino acid will break down in a process called denaturation. A base will neutralize an acid. Bases taste bitter and will feel soapy in the mouth. Household bases include ammonia, bleach, alcohol, drain cleaner, and laundry detergent. Bases work well for cleaning because they strip, or accept protons into their formula, this is good for getting rid of stains or streaks on a window or mirror. Bleach is a very strong base, so strong in fact, that it will strip the pigment out of most clothes!

Solution terms

An acidic solution is any solution that has acidic

properties like corrosion of metal or sour taste.

A basic solution is anything that has alkaline properties like stripping color, or material or having a bitter soapy taste.

Strength of acids

A strong acid contains more ions and is more conductive while a weak acid has less ions making it less conductive. The stronger the acid, the more ions it contains, which directly relates to the conductivity. Acid dissociation measures the strength of acids by how quickly the acid will reach equilibrium with a base. The higher the acid dissociation rate, the stronger the acid. A good example of a strong acid would be hydrochloric acid, this acid rates a 0 on the pH scale. A weak acid would be apple juice, rating at a 4

Strengths of bases

The stronger the base, the less ions, meaning that it will conduct less and vise-versa. A strong base has more electrons, so it is more negative than an acid. (also

why strong bases rip away at certain chemicals) whereas a weak base will be more lacking in electrons, and more "reluctant" per say to "steal" protons.
One example of a strong base would be paint stripper. Paint stripper can literally <u>strip</u> the electrons from the paint, making it lose its chemical composure.

What is pH?

PH stands for "potential of hydrogen" as the chemicals climb the scale, they become greater in basic strength and namely, *potential of hydrogen.*
As chemicals fall in the scale, they become more apt to give

away these negative electrons, so they are acids, meaning that they have pretty much "given up" their *potential of hydrogen.*

What is pOH?

pOH is a measure of alkalinity in solutions. It measures hydroxide ions in a solution.

pH ranges
The pH range of acids is 6.9 to 0.

Most acids in a household environment will be found in the refrigerator or pantry.

The pH range of a neural substance hovers at around 7.

The most abundant neutral substance is water.
The pH range of bases is 7.1 to 14.
Most bases found in a household environment consist of bleach, soap,

laundry detergent, drain cleaner, and paint stripper.

Misc.

A neutralization reaction is a reaction between a relatively equal acid-base relationship, when the acids are done giving their electrons, and the bases are done taking them, the two solutions will have equilibrium with each other.

An indicator is something that shows the pH or pOH of a solution, some of these indicators include litmus paper and phenolphthalein.

A **buffer solution** (more precisely, pH buffer or hydrogen ion buffer) is an aqueous solution consisting of a mixture of a weak acid and its conjugate base, or vice versa. Its pH changes very little when a small amount of strong acid or base is added to it. Buffer solutions are used as a means of keeping pH at a nearly constant value in a wide variety of chemical applications. In nature, there are many systems that use buffering for pH regulation. the bicarbonate_buffering system is used to regulate the pH of blood. You can't hold your breath until you die because when you hold your breath the water buffer in your blood will turn the carbon dioxide into carbonic acid. Meaning that you will eventually cough or gasp for air. The excess in carbonic acid in the lungs will cause an unstoppable urge to cough.

www.ingramcontent.com/pod-product-compliance
Lightning Source LLC
Chambersburg PA
CBHW041945240526
45473CB00033B/609